# LITTLE KIDS
# SOLAR SYSTEM
# COLORING BOOK

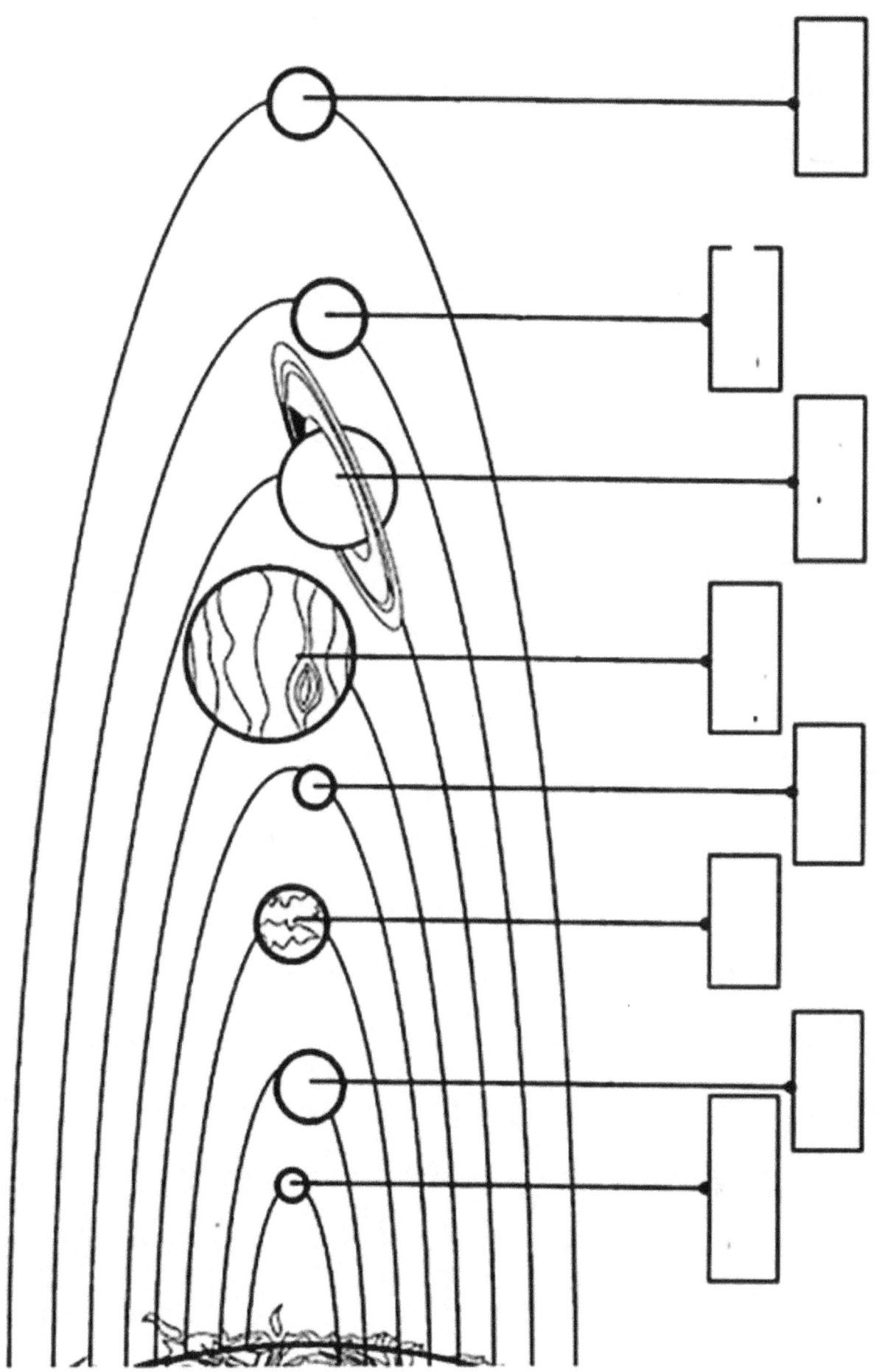

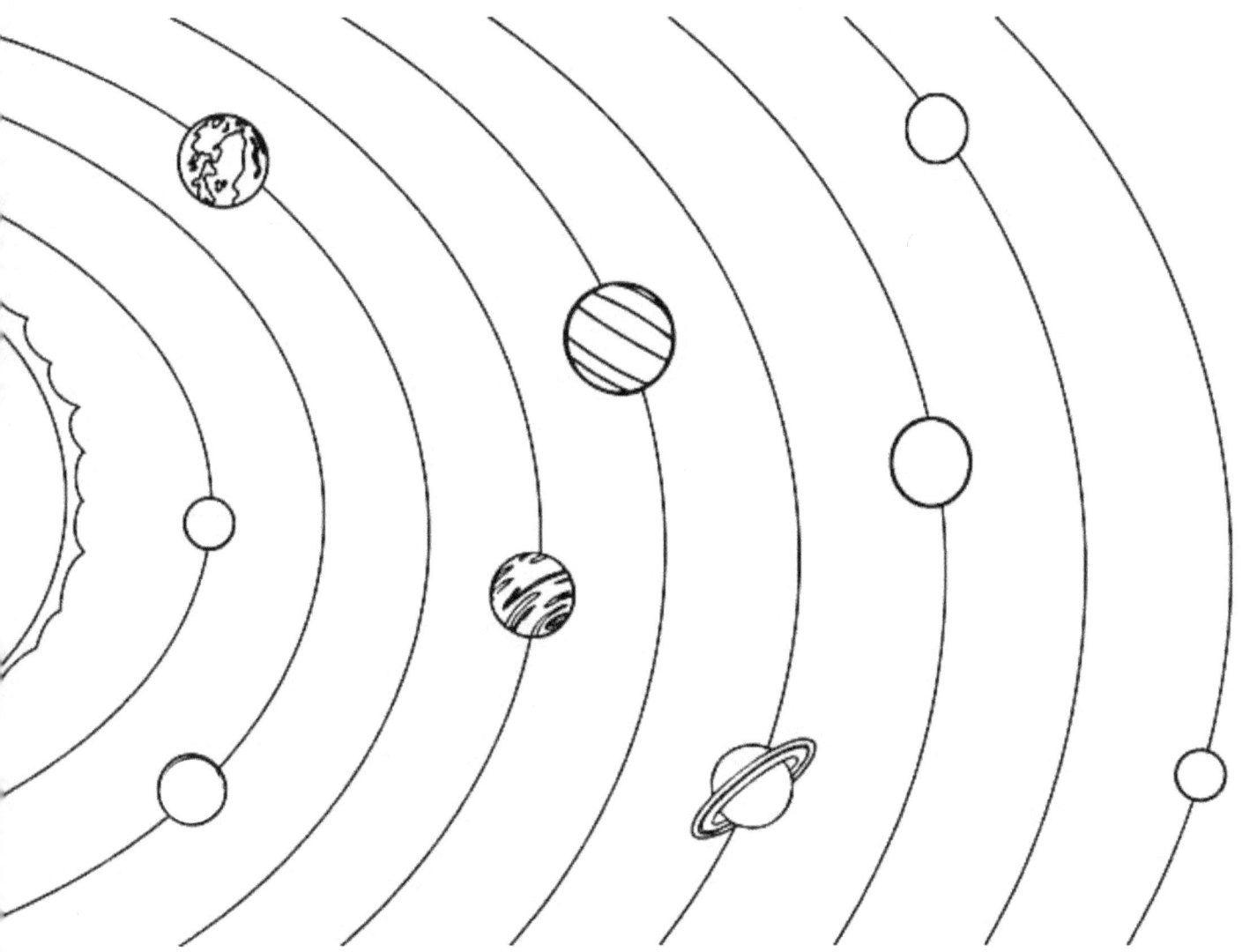

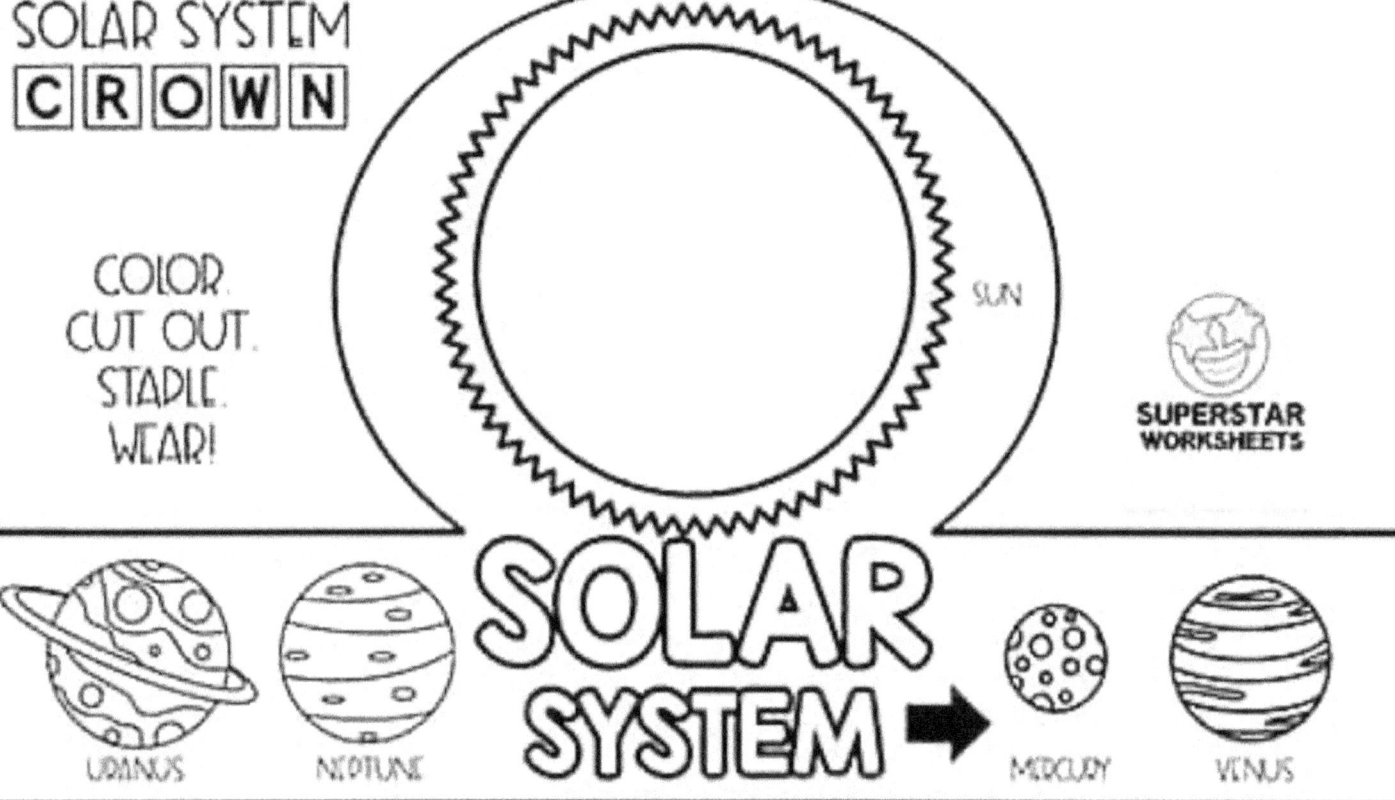

# Saturn

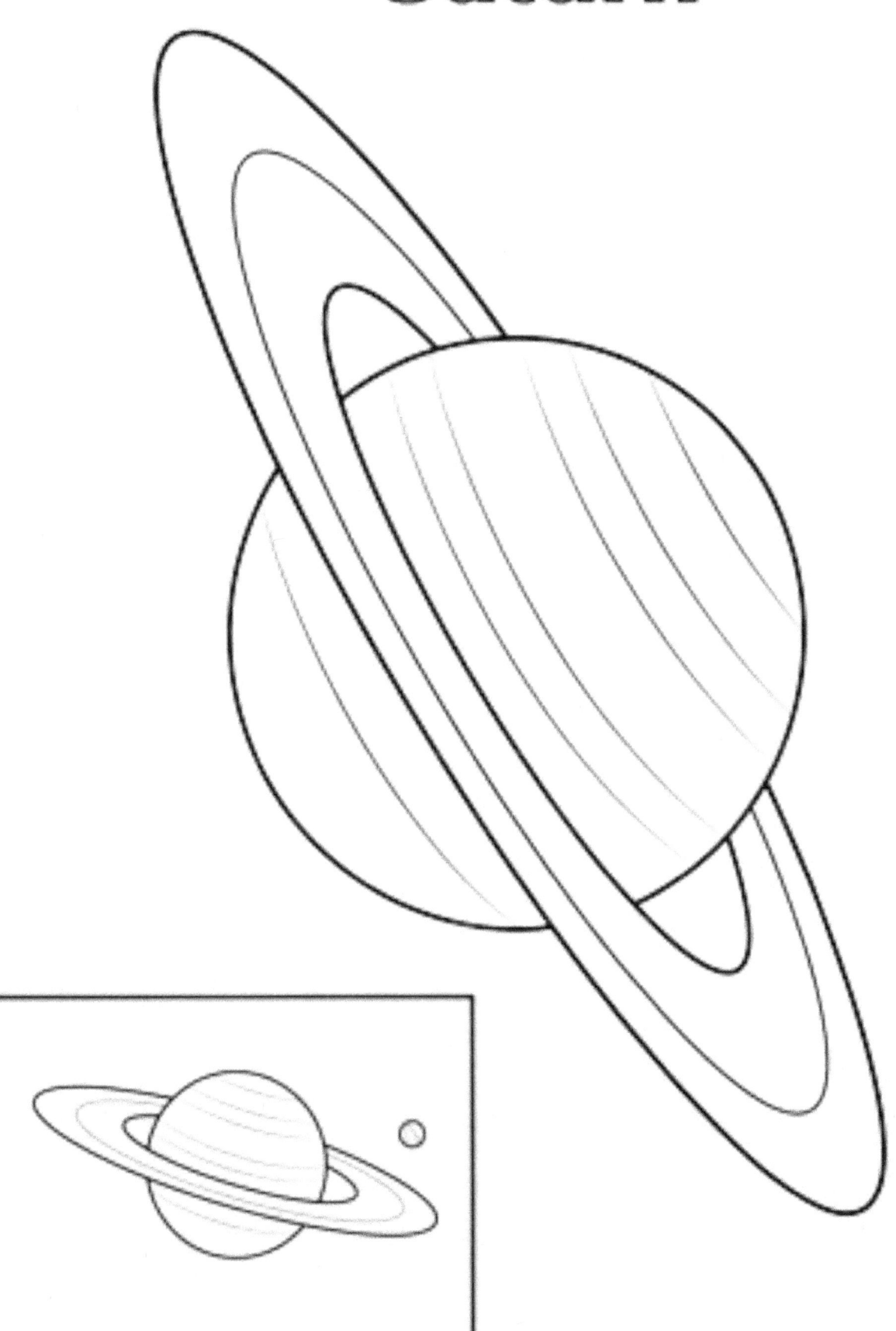

# SOLAR SYSTEM

Mercury
Venus
Earth
Mars
Jupiter
Saturn
Uranus
Neptune

_____
_____

Mercury         Venus         Earth

  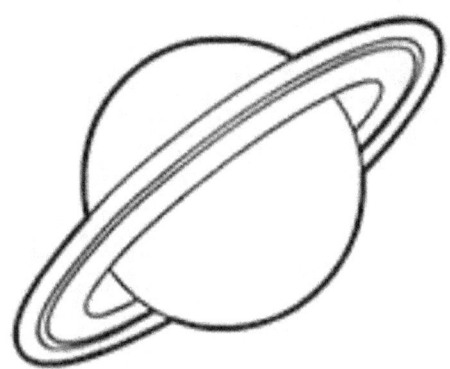

Mars         Jupiter         Saturn

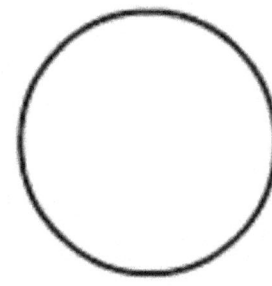 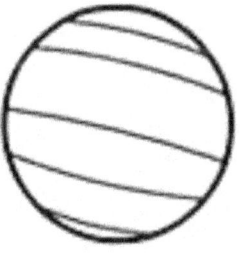 

Uranus         Neptune         Pluto

# Planetele sistemului solar

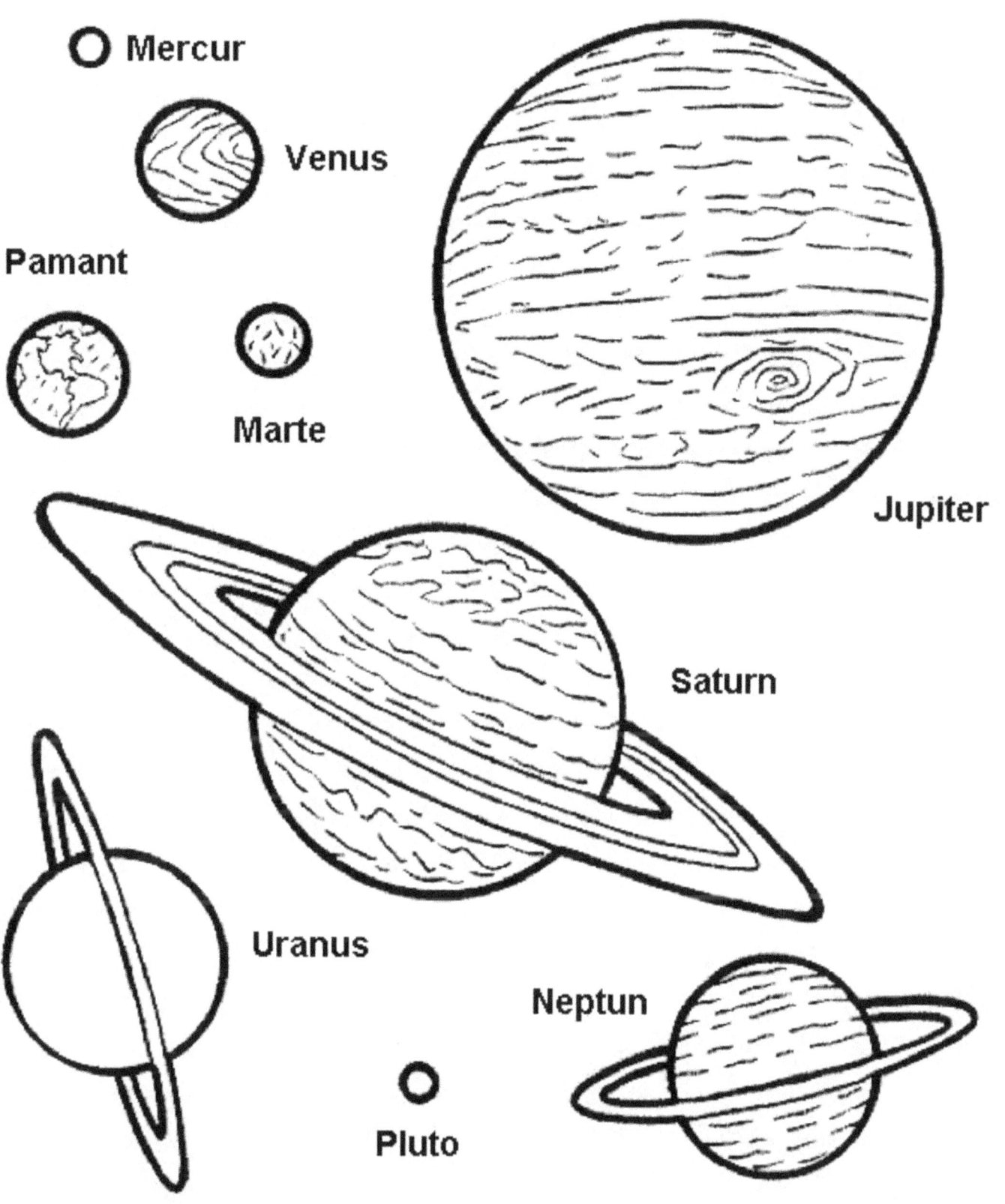

Mercury

Venus

Earth

Mars

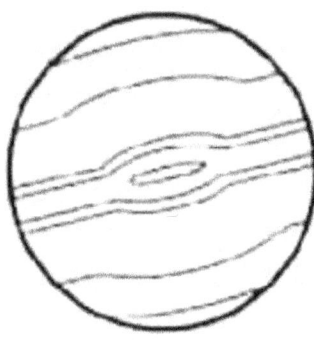
Jupiter

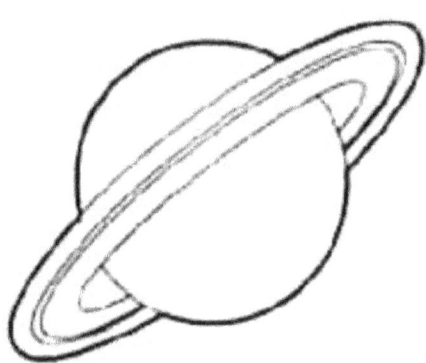
Saturn

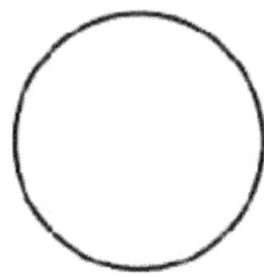
Uranus

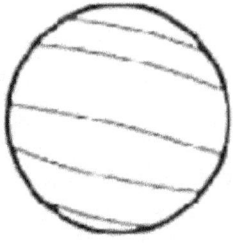
Neptune

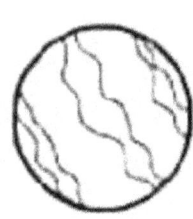
Pluto

# PLANETS

Mercury
Venus
Jupiter
Earth
Mars
Saturn
Uranus
Neptune

# Planet Saturn..!

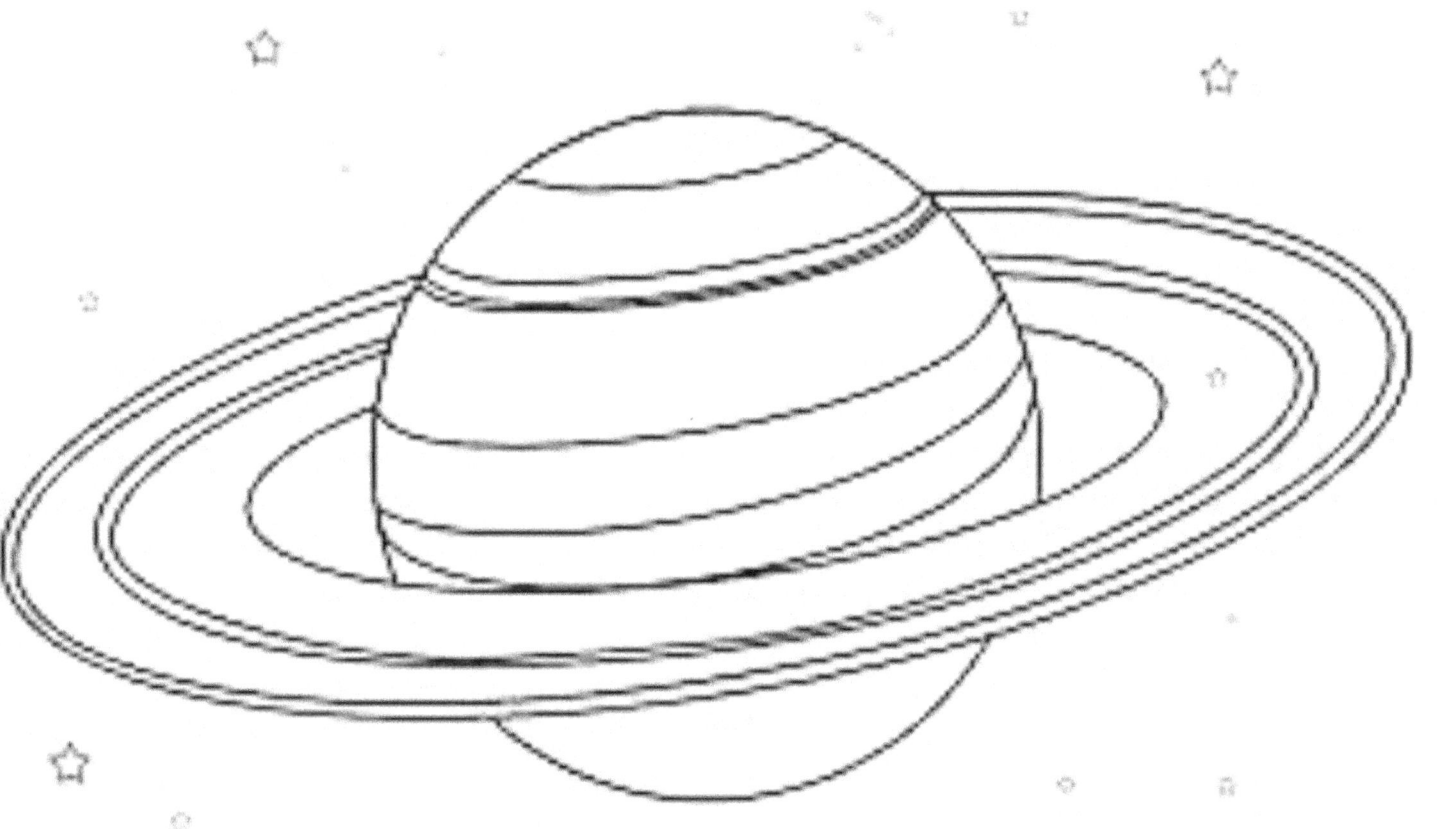

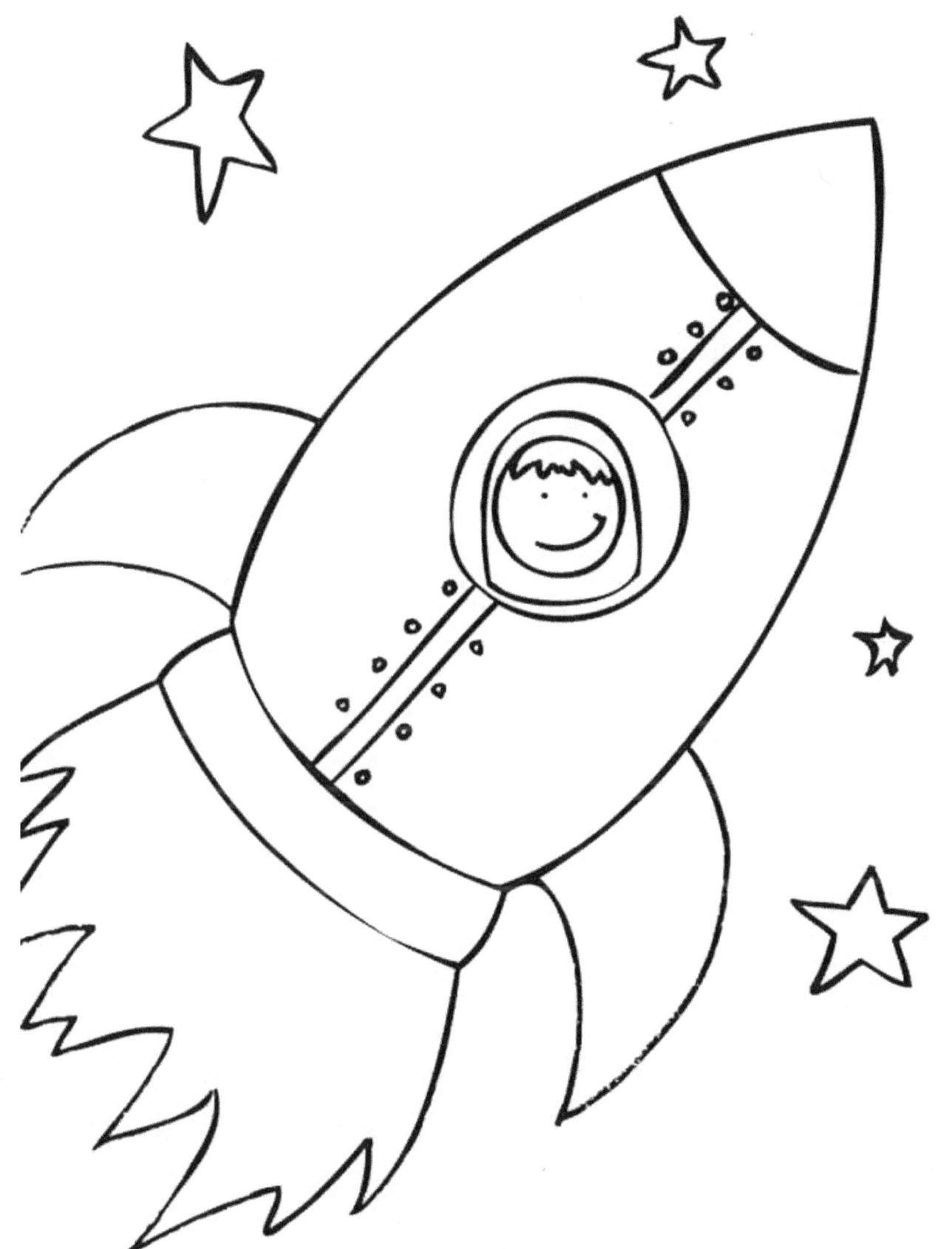

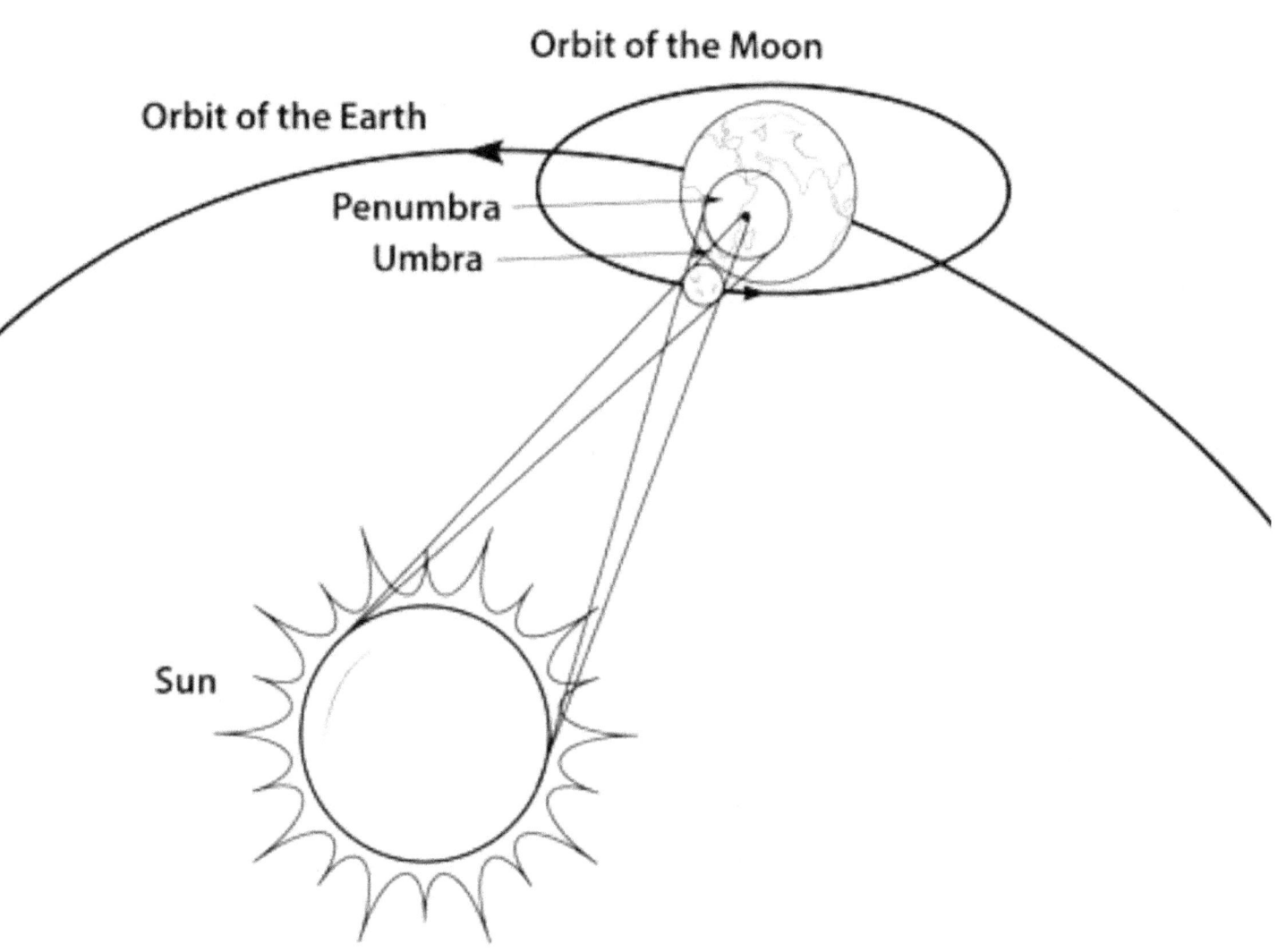

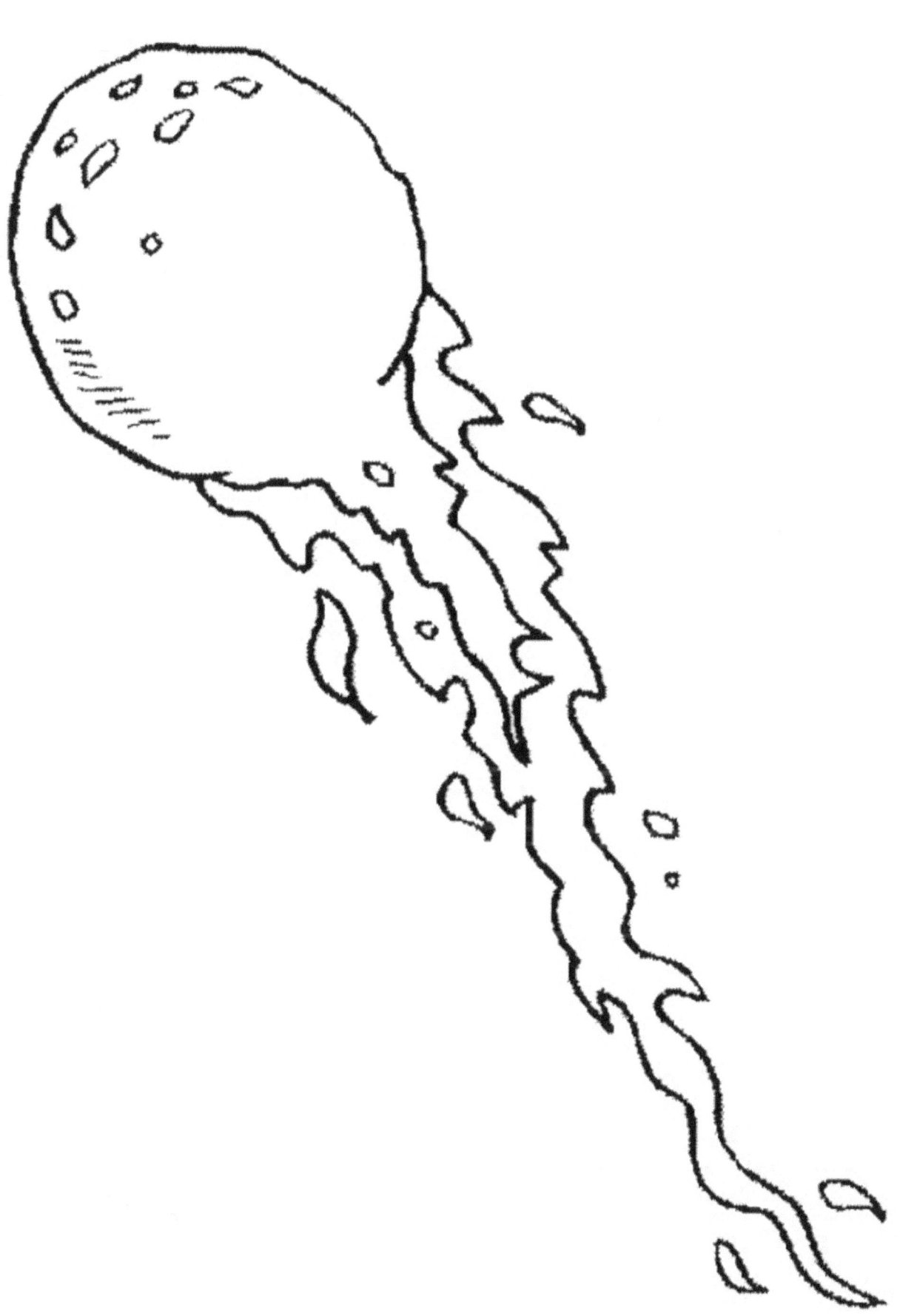

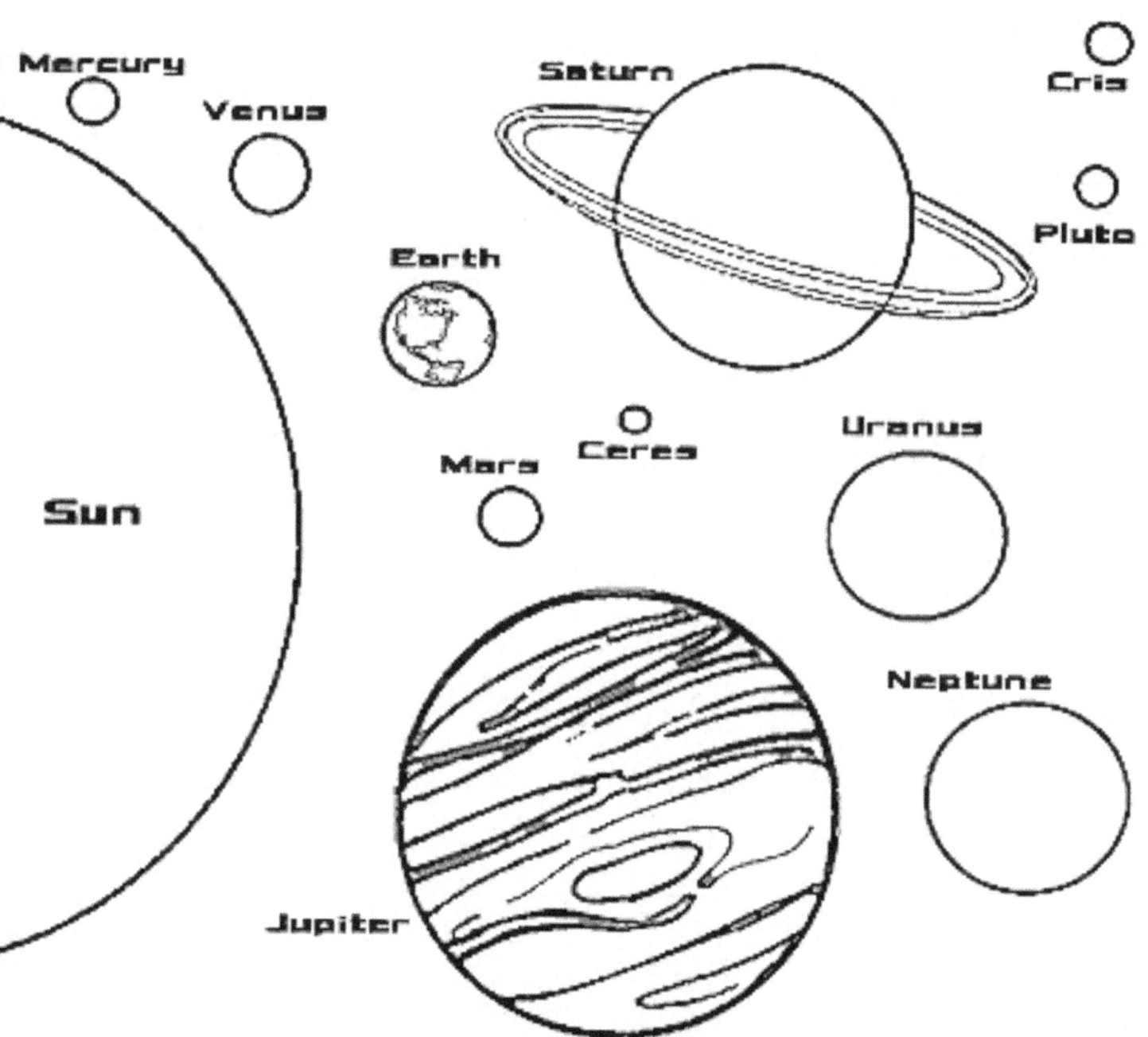

# Solar System Planets.

www.ingramcontent.com/pod-product-compliance
Lightning Source LLC
Chambersburg PA
CBHW081446220526
45466CB00008B/2529